Ford Sierra Cosworth

CLASSICS IN COLOUR

Ford Sierra Cosworth

COLOUR, DATA AND DETAIL ON
PRODUCTION CARS,
SPECIALS AND SPORT

**Dennis Foy
and Terry West**

Windrow & Greene Automotive

Published in Great Britain by
Windrow & Greene Ltd
5 Gerrard Street
London W1V 7LJ

© Windrow & Greene Automotive, 1992

A C.I.P. catalogue record for this book is available from the British Library.

ISBN 1 87200 496 2

Book Design: *ghk* DESIGN, Chiswick, London

Printed in Singapore

Contents

Ian Wagstaff

Acknowledgements

Whilst the writing of a book is normally a solitary occupation, there are exceptions. The press launches of the four Sierra Cosworth production models were gregarious affairs and we would like to thank Ford Motor Company and Ford of Europe's Public Affairs teams for the opportunity to get to know each new car ahead of its on-sale date in a convivial atmosphere.

The inclusion of several special feature cars in the book is likewise the result of good working relationships with a variety of people in and around the industry. Our thanks are extended to Jim Carter, Frank Cirino, Sue Collins, Amanda Evans, Graham Goode, Timothy Lambert, Mike Landsell, Gordon Lewis, Mark Lovell, Tony Mannock, Alastair Mayne, David Power and Dick Prior; in each case, the thanks extend to their respective families, colleagues and employees.

Then there are those who smooth the way to getting trackside access on race and rally events; singled out for thanks are the press teams at the RAC Motorsports Association, Brands Hatch Leisure, Castle Combe, Mallory Park and Silverstone circuits.

The feature cars were shot by Dennis Foy using Bronica equipment and the motorsport action by Terry West using Canon EOS hardware; in both cases Fuji film stock was used. Additional material has been furnished by Ian Wagstaff and by Ian Kuah and the Ford Photographic Unit — latter shots are from the Performance Ford *archives.*

Final — and special — thanks go to our respective wives, Pat and Jackie, and to other members of our families for their patience and understanding whilst we went out testing and photographing the cars drawn together in this, our first joint venture.

Dennis Foy and Terry West
May 1992

Introduction

A massive whaletail spoiler gave the game away: this was no ordinary Sierra.

Originally developed during the late seventies as a replacement for the much-loved (although nobody can say quite *why* it was loved) Cortina, the Sierra had met with an underwhelming response. Its 'jellymould' looks were viewed as strange, its high-speed motorway cruising stability was distinctly suspect, and its cornering abilities were dreadful. Faced with such a negative reaction to their multimillion-dollar investment, Ford's engineers had to clean up the Sierra's reputation and clean it up fast.

The original RS Cosworth was launched to the press in December 1985 — even today, hard-nosed motoring journalists remember the car fondly. (Ian Kuah)

The first of the sporting Sierras came in 1983, when the XR4i arrived. This was, in fairness, already under development when the main range went on sale in 1982, but there was now a positive push to get this high performer out on to the streets. Then, as now, whilst not everybody shopping for a new car wanted to own a high-performance example, there was undoubtedly a subtle underlying psychological satisfaction in at least knowing that there was a faster and more powerful variant available. Reaction to the XR4i was mixed, but it successfully paved the way for a new breed of car — the supersaloon.

It was essentially a twist of fate which brought about the whale-tailed Cosworth-engined Sierra which took the world by storm in 1986. Some three years previously, Stuart Turner had ascended to the rank of Director, Ford Motorsport Europe, and one of his first actions had been to cancel two programmes: the C100 endurance race car and, more significantly, the Escort RS1700T rally machine. The reason for cancelling the latter project was the belief that to create what was essentially a MkII Escort concept (albeit taken to new technical extremes) within a MkIII front-wheel-drive bodyshell was tantamount to fraudulent practice. This, however, left a gaping hole in the company's motorsport programme. What was needed, declared the astute

Mr Turner, was a machine far more directly related to the cars which Mr and Mrs Joe Public could go into a showroom and buy.

In the XR4i there was insufficient potential for an event-winning car — its V6 2.8 injection engine would never produce more than about 220 brake horsepower, when in excess of 300 horses was needed — and the front-wheel-drive cars in the Ford stable, Escort and Fiesta, were wholly unsuited to serious rallying programmes because of their inherent lack of traction. But if a lighter, more tuneable engine could be inserted into the Sierra bodyshell, then the potential was there to produce a car which would not only be a successful rally weapon but also a highly worthwhile racing prospect. At that point, interestingly enough, the primary rally assault was to be headed by a completely separate car, the RS200. Fate was to intervene, however, during the final days of preparation for what was to become the Sierra RS Cosworth — at that point being groomed as a race-track machine —forcing it also to be competitive in rallying.

Following a number of serious incidents and accidents, FISA, the international governing body of motorsport, outlawed the Group B class of rally car, leaving only Group A and Group N models; in both cases the remaining classes had to be essentially the same as cars on sale in showrooms, and cars of which at least 5,000 were built for sale. With the only-just-becoming-a-success RS200 no longer legal, Ford Motorsport at Boreham, Essex, suddenly had to shift the emphasis on to the Sierra RS Cosworth for its rallying activities.

That gigantic whaletail spoiler was deemed essential for rear downforce. The car was conceived as a racer and then road legalised. (Ian Kuah)

But we are jumping the gun slightly. Perhaps this is an apposite moment to explain just how much political manoeuvering took place behind the scenes in 1983 and explain how the Sierra RS Cosworth came into being.

It all started when Stuart Turner persuaded Walter Hayes (then Vice-President of Public Affairs at Ford's world headquarters in Dearborn, but still at heart a motorsport man) to accompany him to a Touring Car race meeting. Hayes was dismayed to see the then Ford combatant, the Capri, being well and truly trounced by the full-house SD1 Rovers. (The Capri of Gordon Spice had peaked in 1980 and the big Rovers had started to come good a year later.) Hayes' initial embarrassment soon turned to positive fury, and the sanction to press ahead with a development of the XR4i's American cousin, the XR4Ti Merkur, was given forthwith. At the same time, Hayes authorised a working programme which would mesh together a variety of components into a new breed of Sierra for European use.

Very shortly after that race meeting, and with a touch of inside knowledge, Turner took a couple of Ford's senior directors to look over the Cosworth Engineering facility. In the course of that visit, they 'happened' to see an all-new dual overhead camshaft, sixteen-valve cylinder head which had been developed for the Pinto engine — the very engine which propelled the vast majority of Sierras. An informal lunch followed, at which it was suggested that this might just be the key to the new Sierra project's motorsport role. The Ford of Europe directors fell for the ploy hook, line and sinker, and the project acquired two more heavyweight allies.

Just as the public was getting used to the sight of RS Sierra Cosworths, along came the killer: the 1986 evolution-special RS500.

With Those Who Mattered on his side, Turner was able to call a formal meeting, attended not only by his top brass within Motorsport but also by Rod Mansfield of Ford Special Vehicle Engineering. Peter Ashcroft had already determined that in order to be competitive the car would need to be turbocharged, and had established that Cosworth Engineering considered the project viable. Rod Mansfield's task was to make sure that the required 5,000 examples of the new car would be produced with the minimum of hiccough and delay and without being unduly compromised in terms of race- and rally-winning potential. Being a hot-rodder at heart, Mansfield ensured that the project remained unsullied by the wishes of the Marketing, Service and Warranty people and followed the initial brief of being a road-legal racer.

Although the car needed to be a two-door hatchback in the interest of minimising kerbweight, there was much resistance to the notion of using the 'six lite' shell already employed on the XR4i. This car was only just being accepted for what it was, a grand tourer rather than a sports saloon, and nobody wanted the new project to be perceived as a variation on the same theme. In 1983 a new two-door bodyshell had been introduced for the Sierra, having a conventional rear side window, rather than the split affair of the 4i; this, agreed everybody engaged in the project, would fit the bill perfectly.

As the car was to be for race-track use, and as effective use of downforce is a pre-requisite of winning races in 150+mph machines,

A second spoiler — taken from the XR4x4 Sierra — augmented the whaletail of the car from which the RS500 was evolved. This gave yet more downforce.

it was decided that the car would need radical spoilers front and rear: too little downforce would result in high-speed instability. Given that Touring Car race rules require that the silhouettes of both road and race versions are identical — as must also be such features as air intakes — then it followed that the wings and fins needed on the track cars would also have to feature on the road-going versions.

Accordingly, the car which finally appeared in public was as far removed from the Sierra 1.6L of 1982 as it possibly could be. It had a deep front airdam with slotted spoiler, a pair of broad air-extractors set into the bonnet, and what can only be described as a huge rear spoiler hanging from the tailgate. If the car looked like a street-racer, that is precisely what it was.

Following Peter Ashcroft's suggestion that Cosworth should look into turbocharging its new cylinder head development, the company went further still: it created an entirely new engine which, while still using the Pinto cylinder block in its 2.0-litre form, was appreciably different from the Ford product. Not the least of these differences was the fact that it produced a nett horsepower figure of 204 and a matching level of torque. Known as the YBB, the engine's power output was achieved by driving charge-cooled air into the chambers via a Garrett AIResearch T.03 turbo-charger. Backing up the engine was a beefy clutch, leading the power in turn to a tough gearbox. As none of Ford of Europe's existing transmissions were able to withstand the torque loadings brought about by the new Cosworth engine, the international network of suppliers was trawled and a suitable 'box found within the American Mustang. Built by Borg-Warner, the T5 was as hard as they come and a perfect choice for the newcomer.

Under the tail of the car lived a suitably beefy differential unit, its substantial final drive gear and pinion controlled by an FF-developed viscous coupling to ensure that traction would be divided equally between the two rear wheels. The fully independent suspension system, though obviously related to the lesser Sierras, was much uprated in terms of both springing and strength. A disc brake featured on each corner of the car, and these were prevented from being over-zealous in their stopping power by a Teves electronic anti-locking system. Criss-cross aluminium alloy wheels, forged by Ronal of Germany, were surrounded by Dunlop D40 tyres in a size of 205/

50/15. As with most Rallye Sport Fords, the team responsible for designing the car in road trim had favoured a specific make and model of tyre. (They usually managed to persuade the cost accountants that multi-sourcing of components was *not* a good idea. Not always, as owners of the MkII Escort RS Turbo will attest, but usually!)

A reasonable degree of cabin civility had been ordered by the marketing department and the package put together by Rod Mansfield's team at SVE fitted the bill nicely. A pair of reclining Recaro seats, with height adjustability for that of the driver, started the ball rolling, and a deep-pile carpet added to the quality feel of the interior. The high-level fascia from the Sierra range was adopted for use in the RS Cosworth and adapted for this specific application by including a boost gauge in the corner of the tachometer. This was not a case of tooling up specially for the newcomer, but rather of raiding the parts bin: the same tachometer was already to be found gracing the instrument panel of the XR4Ti Merkur, the US-market model which combined the 2.3 turbocharged unit from the Mustang with the XR4i bodyshell.

The RS500 found immediate and enduring success as a racing machine. This is the dominant combination of Andy Rouse and his ICS-sponsored Cosworth in 1987.

Powered window winders for the front occupants (the rear side glazing was fixed) and electrically adjustable and heated mirrors featured among the complement of 'extras' Ford included as standard in their new Sierra flagship, along with a sliding and tilting sunroof, high-grade stereo system, rear screen heater and rear wash-wipe mechanism. The cabin was as civilised as the exterior appearance was extrovert.

The car was received in almost rapturous terms by a normally cynical troupe of motoring journalists when first presented to them, just before Christmas 1985, at a pre-launch session in Spain. Superlatives flowed like the poolside Sangrias, while the nearest thing to a criticism was a concern that the massive rear wing reduced rearwards visibility. In every other respect the 150mph super-saloon met with fulsome praise and Ford were spoilt for choice in the number of compliments they were able to reproduce in their sales brochure.

Immediately after the fuss had died down, the public were treated to the ace up Ford's corporate sleeve; the RS500.

The rules in force for international motorsport allowed

In contrast to the three-door car, the first saloon Sierra Sapphire Cosworth was a subdued machine — at least in terms of appearance.

manufacturers to evolve a proportion — 10 percent — of their production run into a yet faster and more serious machine, and this was the case with the RS500. Power was increased to 224bhp, but not merely by altering the amount of boost being produced by the turbocharger: on the contrary, the boost was actually slightly lower than that of the car from which the RS500 had evolved — but more air was being delivered because the turbocharger had been increased in size and many of the moving parts within the engine had been upgraded accordingly. The idea of the evolution engine was to allow something like five hundred brake horsepower to be produced — an awesome amount of muscle from a two-litre engine. The evolution process also included further aerodynamic developments (an additional spoiler, the one found on the XR4x4 Sierra, was added at the rear and the front spoiler was further extended downwards), while there were less obvious but equally important changes made to the suspension system.

The RS500 did the trick, as in 1988 Andy Rouse took one to a class win in the British Touring Car Championships. He repeated the feat a year later, and Ford achieved the hat-trick in 1990 when Robb Gravett won the Championship outright in his Trakstar RS500. Then FISA buckled under pressure from other manufacturers tired of the steamroller effect and outlawed the use of turbocharged cars. It was a goalpost-moving exercise which brought about the instant demise of the Sierra RS Cosworth in saloon car racing. A few privateers made game attempts to build competitive non-turbocharged (atmospheric, in racing parlance) variations on the same theme, but none made a lasting impression.

In rallying, the Cosworth Sierra was never quite as successful. But it was, of course, a rather hasty substitute for the RS200, a car which would surely have gone on to dominate the rallying calendar for several years. As it was, the successes of the Sierra in rallying came no more than infrequently at international level, if only because the other major players in the game were able to devote more time and resources to it. To the outside observer, it seemed that Ford were active on the international scene for no better reason than that they were expected to be there. The two-wheel-drive versions of the Sierra — in particular the brutal RS500 — were unable to deliver their more-than-ample amounts of power as directly as their four-wheel-drive opponents could, and by the time Ford were able to start running a 4x4 version of the Cosworth in international rallying the opposition seemed to be several steps ahead. Even drafting in some first-class drivers failed to give Ford a conclusive advantage, and the period from 1987 to 1991 will not be viewed even through rose-coloured retrospective glasses as a golden age for Ford in international rallying.

In the showrooms, meanwhile, the Sierra RS Cosworth was enjoying a steady period of development. Even while the original streetracer model was being unveiled, plans were underway to produce the next generation, a more subtle model based around the recently-launched Sapphire four-door saloon. This carried over the latest driveline of the Sierra hatchback Cosworth, but was a far more urbane and sophisticated package: appreciably more soundproofing was used in the cabin, the box-sections were injection-filled with expanded polyurethane foam as a means of reducing potential twisting and rattling, and the interior trim was that of the Ghia Sapphire. Only the seats were changed, the Cosworth Sapphire retaining the supportive Recaro recliners which suited a higher performance car than the Ghia. The original RS Cosworth had proved its point and Ford felt that the time was right to clothe its components in a more subtle suit — one which wouldn't turn heads, but which would still deliver the goods when bidden to do so.

That car arrived in April 1988 and was with us only until April 1990. Then, as part of a major overhaul of the entire Sierra/Sapphire range, it gained neutral density rear lights, a 220bhp engine, and four-wheel drive. A development of the much under-rated XR4x4 driveline, the 4x4 Cosworth took the car to new levels of grip and roadholding, enhanced the user-friendliness of the package still further (no more tail-slides...), but for some was just a little too tame. One private owner who had sampled the 4x4 described it to the writer as 'like having a bath with socks on!' Nevertheless, in sales terms the car was a success.

At time of writing, the Sierra Cosworth has less than a year to live before the new CDW-27 model of Sierra comes along. There will not be a Cosworth derivative as part of the initial line-up when the 1993 newcomer arrives, but it will follow before long. It will be a 1.8-litre machine, with a turbocharger again developed by Cosworth Engineering. But that is another story, perhaps for a later book...

Whilst some drivers considered the RS Sapphire 4x4 to be a little anaesthetised, others appreciated its awesome levels of traction.

Sierra RS Cosworth Hatchback

Nobody could quite believe it. Ford had taken their mundane (some might even say tedious) Sierra, and by some clever re-engineering and a touch of body cosmetic work had turned it into a serious supercar. News of its approach had been a closely-guarded secret, and when the press landed in Spain in December 1985 the element of surprise was still intact. To this day, the normally jaundiced characters who make up the British motoring press still remember the impact of a car which they would not have thought possible.

Initial driving impressions were favourable, just about everybody being seduced by the sheer animal pleasure of being in a car which would hold five people and luggage, yet would dance through bends and swoop along straights in a manner which had previously been the exclusive right of such brands as Porsche, Ferrari and Lamborghini.

Opposite page: *The purposeful nose of the first Cosworth featured a pair of additional foglamps mounted beneath the bumper line, inboard of the indicator lamps. Lower outboard ducts were to allow cooling air to reach brake discs.*

Below: *As with so many Ford RS products, the RS Cosworth could be bought in only a limited range of colours: white, black or, as pictured here, Moonstone Blue.*

SIERRA RS COSWORTH FACTFILE

Introduced February 1986; last car built December 1986

Bodyshell	Three-door hatchback
Engine type	Turbocharged twin overhead camshaft four-cylinder
Bore x stroke	90.8mm x 77.0mm
Bhp @ rpm	204 @ 6000
Torque peak @ rpm	204 @ 4500
Fuel system	Weber Marelli electronic injection with Garrett AiResearch T.03 turbocharger
Ignition system	Weber electronic
Clutch	240mm diameter self-adjusting single plate
Gearbox type	Borg Warner five-speed manual
Gearbox ratios	(1) 2.95:1 (2) 1.93:1 (3) 1.33:1 (4) 1.00:1 (5) 0.80:1 (R) 2.75:1
Final drive ratio	3.64:1
Drive system	Front engine, rear-wheel drive
Suspension, front	Independent MacPherson struts, lateral control arms, 28mm diameter anti-rollbar
Suspension, rear	Independent semi-trailing arms, inboard coils, monotube dampers, 16mm anti-rollbar
Braking system	283mm x 24mm ventilated front discs, 272mm x 10.5mm solid rear discs. Teves electronic anti-lock mechanism
Wheels & tyres	15" x 7" aluminium cross-wire style wheels, Dunlop D40 205/50VR15 tyres

Wheelbase	2608mm (102.7")	**Length**	4460mm (176")
Width	1920mm (75")	**Height**	1376mm (55")
Weight (dry)	205kg (2651lbs)	**Power/weight ratio**	169bhp/tonne
Acceleration 0-60	6.1 seconds	**Maximum speed**	150mph
Mpg average	21.23mpg	**Price new**	£16,500

Left: Cross-wire pattern wheels were used on the first generation of Cosworth. These were not particularly easy to clean, proving themselves prone to accumulating dust from the (often overworking) braking system. Centre spinner concealed hub studs.

Yet those early examples were not perfect. The front steering geometry, which had been carefully selected to endow the chassis with just the right amount of steering feedback, was prone to knocking out, and a number of owners of those early cars grew tired of constantly having to get the tracking reset. It would take only as simple a matter as mounting a kerb when parking to upset the car's balance.

There were also those who felt distinctly vulnerable in a car with such restricted rear visibility. The author's own observation was that it was only just possible to see the blue light on the roof of the police car immediately behind — and there always was one. The bewinged monster seemed to attract the attentions of the local constabulary in the same way that white sharks are drawn to a bleeding swimmer off the California coast.

That massive rear spoiler was, we were assured, absolutely essential to control the car's stability at high speeds; had it not been there, the car would have wandered from its intended line when cruising on the *autobahnen* of West Germany at speeds in excess of 150kph. It was inconceivable, of course, that British motorists would ever exceed the 70mph national motorway speed limit...

From the inside, there was little to distinguish the car from lesser Sierras. Only a pair of Recaro front seats and a smaller steering wheel (which all but concealed the boost gauge tucked into the corner of the tachometer) gave any clues as to the special nature of the car. From the outside, however, it was a different matter.

As well as that huge whaletail spoiler — which some people interpreted as a bit of a mickey-take of Porsche's similarly-equipped 911 Turbo — there was also a set of four arch extension pieces, a pair of wider rocker panel covers, and a totally revised nose treatment which featured a proper radiator air intake and a deep front spoiler. The finishing touch was a pair of air vents let into the bonnet; whilst these were intended to release excess hot air (of which the turbocharger unit generated plenty), they were not, for purely visual reasons, ideally placed: had function taken priority over form, they would have been sited further up the bonnet and closer together.

Those wheelarch extensions were more than cosmetic, being needed to cover the 15" x 7" criss-cross Rial aluminium alloy rims which were standard issue on the model — and these in turn were necessary to allow the use of a set of substantial 285mm diameter, 24mm thick ventilated disc brakes with four-pot calipers to squeeze

Above: Bonnet louvres were deemed necessary to allow hot air to evacuate the engine bay. Under-bonnet heat was always a problem on Sierra Cosworths, thanks to the amount of energy generated by the turbocharged 2.0-litre engine.

their anti-fade pads. Discs also featured at the rear, smaller in diameter and non-ventilated. Anti-locking technology was a standard Cosworth feature, this being a further development of the Teves-designed system already found on the previous year's Scorpio. Electronically activated, the ATE microprocessor momentarily relieved the pressure on the discs, as soon as it showed signs of locking up.

The suspension arrangement of the Cosworth was a fine-tuned development of that already in service beneath lesser Sierras, albeit with one or two vital additions such as a rear anti-roll bar of 14mm diameter. The front bar was of a massive 28mm diameter and worked in conjunction with special coil springs and F&S gas-filled damper units. The result was closely-controlled body roll allied to a surprisingly supple ride quality.

A fast-action steering rack with variable-rate power assistance (the slower the car was moving, the more assistance the driver gained) allowed rapid and well-balanced changes of direction without the driver needing to be in the Schwarzenegger class of muscular development. The precise degree of assistance the driver would gain had been something of a problem for the development team at Ford SVE to determine: pre-launch cars were, according to Rod Mansfield, something of a handful because they were a mite twitchy. By the time the car went into production, it was 'de-sensitised' and thus more controllable. A few of the early production models were still twitchy., but by the time the car had run into four figures on its production schedule the system had settled down to its final settings.

Not all the changes wrought on the standard Sierra were necessary for road-going Cosworth-engined models; quite a number of the special suspension components were designed for interchange-ability with a package of special racewear items which were being developed alongside the road car.

The RS Cosworth in its original form was with us only until July 1986, when the RS500 came in. Having said that, there is at least one example known to the author registered in August 1987, and there are rumours (but then there always are...) of a handful sitting in a warehouse somewhere in the south of England which have yet to be given registration numbers.

If the rumour is true the owners can hardly be blamed, for in making those 5,000 examples of streetfighter Ford produced an instant classic which will be fondly remembered for many years to come. In that respect, the Cossie is in the same league as the Jaguar E-Type and others of that coveted ilk.

Above: *Although the original car was a race-car in thin disguise, few private owners ever took their machines on to the circuits. However, privately-run test days (often by tuning concern Collins Engineering) allowed drivers to make the occasional forays on to tracks.*

Sierra RS500 Cosworth

To the purists, and to anyone occasionally taken with the urge to go out and tear holes through the traffic laws, this was the ultimate Cosworth. Conceived to have fangs in place of the normal teeth of its immediate predecessor, the RS500 was created for one purpose and one purpose only — to win races.

This was a car for the Top Guns — for those who actually fancy having a British Aerospace piece of warmongering hardware, but who lack the pilot's licence, the £15 million, or both. The RS500 in full cry was a pretty good substitute.

Much of its appeal to owners was the way the power kicked in. Whereas the original Cosworth was pleasingly progressive, the RS500's turbocharger installation saw to it that forward motion took on an all-or-nothing attitude: the big T.04 turbocharger (originally designed for truck engines) didn't do anything until the engine was spinning at perhaps 3500rpm, but would at that point wake up with a vengeance and transform the car from pussycat to tiger. On a good day, vision through the windscreen would become a blur — rather like watching a video in fast-forward mode.

Alongside the engine was a duplicated fuel rail system, enabling twice the amount of fuel to be fed into the cylinders after a minor degree of alteration; in roadgoing form the second bank of injectors was de-activated and the car ran on the same four primary injectors as its less powerful predecessor. Quite understandably, a number of owners immediately rushed to their nearest speed shop and had the second rail opened up — often with disastrous results, for although more fuel ought to make more power, it was never that simple.

Only 500 examples were built, these evolving from the standard Cosworth in accordance with the rules then in existence for international motorsport. The whole point of that second rail of fuel injectors, the mammoth turbine unit and the various other goodies incorporated into the stronger YBD

cylinder-block (such as much-improved lubrication and re-designed head porting) was to allow the extraction of anything up to 500 brake horsepower, and sure enough the car was immediately seized upon by a number of Britain's premier tuning companies. Mountune, Brodie Brittain Racing, Andy Rouse Engineering and Graham Goode Racing were just some of those who took the car on board, up-graded everything for track use, and started to win events with it.

During 1988, 1989 and 1990 the car was unbeatable, dominating not just the British Touring Car Championships but also the corresponding German and

Detecting A Genuine RS500

In view of the number of RS Cosworth three-door hatchbacks which have been converted to RS500 specification, or at least visual impersonations of an RS500, the scope for a potential owner to be duped into buying a non-genuine example is great. There were in total 392 black cars built at Bedworth, with the balance of the run being 52 each in Moonstone Blue and in white.

According to Tickford, the VIN's (Vehicle Identification Numbers) of the RS500s they produced were from -GBBEGG38600 to -GBBEGG39099, and this should be the first check as to any RS500's authenticity. All gained a larger intercooler, eight injectors on two rails and a T.04 turbocharger, and had engines with the prefix YBD.

As well as the 496 accounted for above, there were four further white examples which were development cars. They were numbered -721, -750, -759 and -794. They are understood to have been destroyed, but might possibly still exist — one never really knows with Ford!

Opposite page: The RS500 was outlawed from motorsport following complaints from other manufacturers who couldn't match the car's awesome power levels. As this 1989 shot of the startline at Silverstone shows, they did have a case!

Above: The additional tailgate spoiler was an off-the-shelf part, taken from the Sierra XR4x4. This gave a further enhancement to downforce — essential at racing speeds, which could reach as high as 180mph. Dedicated graphics were another feature.

Below: As well as the foglamp deletion, the RS500 nose was also distinguishable from its predecessor by the use of an additional rubber spoiler (an air splitter, which directed air around the car rather than allowing it to go under the floorpan) and by dedicated wing graphics.

SIERRA RS500 COSWORTH FACTFILE

Introduced August 1987; built at Tickford from set-aside standard RS Cosworths taken from stock; most deliveries made by October 1987

Bodyshell	Three-door hatchback
Engine type	Turbocharged dual overhead camshaft four-cylinder
Bore x stroke	90.8mm x 77.0mm
Bhp @ rpm	224 @ 6000 **Torque peak @ rpm** 206 @ 4500
Fuel system	Weber Marelli electronic injection, with eight injectors and Garrett AiResearch T.04 turbocharger
Ignition system	Weber electronic
Clutch	240mm diameter self-adjusting single plate
Gearbox type	Borg Warner five-speed manual
Gearbox ratios	(1) 2.95:1 (2) 1.93:1 (3) 1.33:1 (4) 1.00:1 (5) 0.80:1 (R) 2.75: 1
Final drive ratio	3.64:1
Drive system	Front engine, rear-wheel drive
Suspension, front	Independent McPherson struts, lateral control arms, 28mm anti-rollbar
Suspension, rear	Independent semi-trailing arms, inboard coils, monotube dampers, 16mm anti-rollbar. Provision for racing arms to be fitted in place of standard items
Braking system	283mm x 24mm ventilated front discs, 272mm x 10.5mm solid rear discs, Teves electronic anti-lock mechanism
Wheels & tyres	15" x 7" aluminium alloy cross-wire style wheels, Dunlop D40 205/50VR15 tyres

Wheelbase	2608mm (102.7")	**Length**	4460mm (176")
Width	1920mm (75")	**Height**	1376mm (55")
Weight (dry)	1219kg (2682lbs)	**Power/weight ratio**	183.75bhp/tonne
Acceleration 0-60	5.9 seconds	**Maximum speed**	153mph
Mpg average	20.60mpg	**Price new**	£19,995

Above: Quite a few owners changed their RS500s into something a little closer to the racing model. This one belonged to Roger Mayers and was prepared by Collins Performance Engineering. 400-plus horsepower, trick suspension and special Revolution wheels hint at the available performance.

Left: Interior of the RS500 was carried straight over from the RS Sierra Cosworth. Boost gauge in the top left corner of the instrument cluster, a leather-trimmed XR3i wheel and Recaro seats set it apart from lesser Sierras. Note RS logo on fascia insert covering management system.

Australian series. It reached the point that by the end of the 1990 series there had been so many complaints to motorsport's governing body FISA (Federation International de Sporting Automobiles) from rival manufacturers unable even to get near the all-conquering Fords that FISA had no option but to move the goalposts.

Interestingly, the majority of the cars which raced — and there were quite a few of them — were not taken from the stock of 500 which had been painstakingly assembled at the Tickford coachbuilding facility in Bedworth, Coventry. Instead, they tended to be specially-built cars which used a three-door bodyshell from the stockpile at Ford Motorsport and were then completed using parts which had been homologated for racing use as a part of the RS500 evolution programme. It was not necessary for the cars actually to be RS500s — they simply had to conform to the requirements laid down by FISA.

Which meant that most of the 500 Tickford-built cars became road vehicles. A number of these cars were terminally damaged within the first few months of the model's existence and a good few more were stolen, never to be seen again. At present it is estimated that no more than three-fifths of the original run still exists — although there are quite a number of big-wing Cosworths about which have been made to look like the RS500 by having the evolution model's additional aerodynamic aids fitted to the body.

Below: Low, mean and handsome, the RS500 was a car which looked like it was travelling at speed even whilst parked at the kerbside. Unfortunately, it also attracted the attention of thieves. This one was stolen within weeks of being sold as an ex-demonstrator and was never seen again by its rightful owner, glamour photographer Mel Grundy.

Sierra Sapphire RS Cosworth

Strangely enough, the car has never officially been known as the Sapphire RS Cosworth, but simply carried over the badging already seen on the bewinged original hatchback.

Although making no bones about the fact that the car was a development of the three-door version, Ford were keen to point out that the new four-door model was very much a road, rather than race, machine. Much of the driveline was carried straight over from the hatchback, which meant that the engine was the familiar 204bhp powerplant backed up by the same Borg-Warner five-speed manual transmission and 3.64:1 final drive rear axle with viscous coupling.

The bodyshell, however, was totally new. The four-door Sapphire saloon body had been introduced by Ford a year prior to the Cosworth saloon's April 1988 debut and had been well-received by all for its tidy appearance and conventional saloon practicality. It was therefore a logical development for Ford to transplant the Cosworth driveline and suspension arrangement into the new shell — but before they could effectively do this, they had to make one or two minor detail changes.

The first of these was to foam-fill all of the body cavities; this not only deadened the sound, thus making for a more relaxed cruising performance, but also appreciably stiffened the already-good shell. New mouldings were made for the front and rear bumper assemblies, and there were also new mouldings to cover the side-skirts. These

Below: The RS series have only ever been available in manual transmission form. To cater to those who wished to have the acceleration of the RS Cosworth with the ease of automatic transmission, Hartford Motors of Oxford produced a special built-to-order automatic version of the RWD car.

were all further developments of an existing range of add-on panels which were in production under the lucrative RS aftermarket parts banner. A new grille was developed which followed the theme of that seen on the hatchback Cosworth, but which fitted between the widened lamp clusters which were a feature of the Sapphire bodyshell. Those different lamps had also been adopted for the Sierra five-door range at the same time as the Sapphire had arrived, so leaving them in their intended state for the saloon-bodied Cosworth was a necessity.

A subtle bootlid spoiler and a fresh set of aluminium alloy wheels — which were to prove notoriously difficult to clean — were the final exterior touches. On the inside, a level of luxury was decided upon from the start: accordingly, purchasers of the Sapphire Cosworth gained a pair of Recaro front seats, headrests for the rear seats, power windows front and rear, a sunroof, leather detailing for the steering wheel and gearknob, a six-speaker stereo system with radio-cassette

Above: Introduced in 1988, the four-door variant of the Sierra RS Cosworth made use of foam-filling of structural members to stiffen up the bodyshell. Engine was taken straight from the first generation of RS Cosworth and gave out 204bhp —sufficient to give 140+mph performance.

Left: With sufficient plumbing to fill every inch of under-bonnet space, the Sapphire Cosworth's engine compartment looked crowded. After-market tuners tended to replace much of the Ford rubber hosing with expensive stainless steel-sheathed pipework.

SIERRA SAPPHIRE RS COSWORTH FACTFILE

Introduced April 1988; last car built January 1990

Bodyshell	Four-door saloon
Engine type	Turbocharged dual overhead camshaft, four-cylinder
Bore x stroke	90.8mm x 77.0mm
Bhp @ rpm	204 @ 6000
Torque Peak @ rpm	204 @ 4500
Fuel system	Weber Marelli electronic injection, with Garrett AiResearch T.03 turbocharger
Ignition system	Weber electronic
Clutch	240mm diameter self-adjusting single plate
Gearbox type	Borg Warner five-speed manual
Gearbox ratios	(1) 2.95:1 (2) 1.93:1 (3) 1.33:1 (4) 1.00:1 (5) 0.80:1 (R) 2.75:1
Final drive ratio	3.64:1
Drive system	Front engine, rear-wheel drive
Suspension, front	Independent McPherson struts, lateral control arms, 28mm anti-rollbar
Suspension, rear	Independent semi-trailing arms, inboard coils, monotube dampers, 16mm anti-rollbar
Braking system	283mm x 24mm ventilated front discs, 272mm x 10.5mm solid rear discs, Teves electronic anti-lock mechanism
Wheels & tyres	15" x 7" aluminium alloy radial spoke style wheels, Dunlop D40 205/50VR15 tyres

Wheelbase	2608mm (102.7")	**Length**	4494mm (177")
Width	1920mm (75")	**Height**	1376mm (55")
Weight (dry)	1251kg (2754lbs)	**Power/weight ratio**	163bhp/tonne
Acceleration 0-60	6.09 seconds	**Maximum speed**	145mph
Mpg Average	23mpg	**Price new**	£19,000

Above: *Keen to keep the car's profile as low as possible, Ford dispensed with the bonnet louvres. This proved an unwise move, as quite a number of owners reported problems with under-bonnet heat build-up when caught in town traffic; there is a healthy trade in after-market vent installations.*

Above: These wheels were new for the Sapphire Cosworth — and, from a cleaning viewpoint, proved the worst ever designed. Their many nooks and crannies accumulated deposits from the brake disc pads which would become permanently ingrained on the sub-surface of the aluminium alloy.

unit and separate amplifier, graphic display module warning of doors open or lights failing to work, heated front screen, heated and electrically-adjustable door mirrors, and centre console with cassette tape storage.

As with the hatchback which had preceded it, the car was looked upon very favourably by all who tried or tested it. Its blend of low-key appearance, practicality and overall urbanity saw to it that the model did not attract too much undesirable attention, whilst the turbocharged 2.0 powerplant provided about as much action as any driver could want or expect from a four-door saloon.

Having said that, the author well remembers being pulled over by one of Cheshire Constabulary's motorway patrol cars about a week after the Sapphire had gone on sale, 'so that we can see what we'll be up against,' according to the pleasant — and informed — officer. (Just before driving away, he commented that the car must be legitimate 'or we'd never have caught you up…')

That parting remark implied that the newcomer would be as vulnerable as its predecessor to the attention of thieves, and regrettably this proved to be the case. Quite a number of cars were stolen from their rightful owners, and this state of affairs seems set to continue. It appears that the only testament from drivers which Ford would not be prepared to use in its advertising would be that from 'motor men', the trade parlance for get-away drivers. They, too, appreciated the low-profile high-performance package which the Sapphire Cosworth provided.

Whilst the Sapphire was never intended to be anything other than a roadgoing machine, there were still those who raced them. Because the car was a development of an existing model there was no need for Ford to seek special homologation, and it thus went on to become a firm favourite amongst saloon car racers. They rarely left their cars in the restricted selection of colours available from the factory, with which road car users had to be content: namely, blue, grey or white.

But restricted colour range or not, the Cosworth in Sapphire form was bound to be at least as successful as its predecessor had been. In fact, it actually gained sales at the expense of other, more established supercar marques, as potential purchasers realised they could have a sensible saloon which could take on the Porsches they had previously favoured — and still have plenty of change too.

Below: *A low-mounted bootlid spoiler — quite a contrast to that of the big-wing RS Cosworth — was a feature of the four-door supercar. Bootlid badging was taken straight from the earlier car and made no reference to the car being a Sapphire.*

Opposite page: The owner of this rear-drive Sapphire Cosworth has made use of the later 4x4 model's air vents as a way of reducing under-bonnet heat in his tuned version of the super-saloon. Surprisingly, the standard wheels have been retained: many owners change to either Revolution or Borbet wheels.

Above: Anatomy of a super-saloon: this Collins Studio cutaway drawing of the Sapphire RS Cosworth shows how neatly everything fitted into the bodyshell of the urbane guerilla. Rear-drive cars are externally identifiable by their amber-lensed front indicators and full-coloured rear lamp clusters.

Below: Leather trim became an option on the Sapphire in 1989, but some owners — such as Kevin Gribble, of Turbo Systems — predated its introduction by having their interiors re-trimmed in hide. This example also features Power Engineering walnut trim fillets on the door cappings and management system cover.

Sierra Sapphire RS Cosworth 4x4

There was a degree of inevitability about the four-wheel-drive variant of Cosworth Sapphire which came into the world during the spring of 1990. The concept of bringing four-wheel-drive executive cars to a wider market platform had been established first by Audi, but Ford had followed suit as soon as possible afterwards — for which read five years — and their XR4x4 and corresponding Ghia estate variant had developed a firm clique of admirers almost from introduction. By the late 1980s, that clique had swelled to quite profitable proportions.

There had, however, been an increasingly strong case put forward for the market to be fed something a little more sophisticated than a twenty year-old cast-iron engine of relatively low power output. The obvious answer was to offer a fusion of the XR4x4 and the Sapphire Cosworth; any work which needed to be done in getting the Sapphire shell to accept the 4x4 driveline could be shared with a number of other developments within the overall Sierra and Sapphire range.

Once again it was Spain — which has a lower density of traffic than most European countries, together with a clement climate —

Opposite page: One of the author's more memorable weekends occurred in 1990, when he took the then-new RS Cosworth 4x4 into the Lake District of northern England. The tenacious chassis made light of all challenges, cutting a tidy path along slippery fellside roads.

Below: Introduced in 1990, the 4x4 model was immediately recognisable by its clear (rather than amber) front indicator lamps, its bonnet louvres and discreet 4x4 badges on the front wings. As is usual for Ford, the pre-launch press evaluation cars were German-built and registered. (Ford of Europe)

car gained neutral density (black, until the lights were on) rear lamp clusters, along with a black decal which joined the two lamps; this visually broadened the car. At the front, the indicator lamps were changed from amber to clear lenses, again deceiving the eye into thinking that the car was wider than it actually was.

Exclusive to the Cosworth was a pair of neat slimline bonnet louvres (these would later appear in the bonnet of the RS Fiesta) intended to overcome the under-bonnet heat problems which the bigger intake charge cooler would bring with it. In all other respects, the bodyshell was the same as that of the car it succeeded.

The devilish-to-clean wheels were carried over from the earlier Sapphire Cosworth (though replaced early in 1992 by some lower-key radial spoked items), but the tyres chosen were new. Rather than the D40 Dunlops, Ford SVE dictated that Japanese-made Bridgestone RE71 hides were the way to go. From the author's experience of pedalling a Cosworth at often indecently high speeds on wet road surfaces, the decision was a sound one. The tyres offer a combination of excellent grip on both wet and dry surfaces with a comfortable ride quality.

Above: Under-bonnet of the 4x4 looks much like any other Cosworth, with the 2.0-litre turbo-charged powerplant taking up more space than might be ex-pected. A downside to the pair of bonnet air ventilation louvre inserts is that they let the rain in as well as the hot air out!

Right: That rear spoiler was apparently able to generate as much downforce at the rear of the Sapphire despite the much smaller dimensions. It is reckoned that the large flat area of the bootlid was a major contributory factor to the overall downforce coefficient.

Rather surprisingly — and perhaps unfairly — in producing the 4x4 variant of RS Cosworth Sapphire, Ford alienated a sector of their previous clientele. The car was, if anything, too safe, too surefooted, and in consequence was considered by some to be less exciting than they would have wished.

At the same time, though, the 4x4 opened up a major new aftermarket industry which has gone on to produce more and more exciting variations. The record at present is in the region of 590bhp, running through an essentially unaltered driveline... Talk of variations, however, leads us into the next section of this book, which is a round-up of some of the more exhilarating and different cars the author has been privileged to test-drive in the course of his normal working life

Opposite page, above: *The enormous number of decals covering the metallic British Racing Green car were needed to help towards the considerable cost of flying it out to the USA two years in a row.*

Opposite page, below: *A battery of additional instrumentation featured in the car. As well as timing equipment and an ultra-accurate electronic compass, there was also a device which warned of the proximity of Highway Patrolmen. A fuel and distance computer was another essential on such a trip.*

Above: *Three team members participated in a 16,000-mile odyssey around America. Needless to say, they couldn't have achieved anything without the help and support of their families. This picture was taken immediately before departure for the 1990 contest.*

The RS2400

This car was the result of a coalition between Power Engineering, the much-respected road and race car preparation company based in Uxbridge and P&P Performance Engineering, the Weston-Super-Mare business headed by top-flight rally star Mark Lovell.

The average turbocharged car — and even the Cosworth qualifies as average in this particular instance — suffers from a syndrome known as 'lag' whereby not much happens at low engine speeds: it is only when the engine comes on to positive boost that the car lights up and really starts moving. Both parties involved in the RS2400 project considered this to be an untenable situation, feeling that overall driveability would be much-improved by redesigning the engine which Cosworth Engineering had already laboured over.

The result was a car with massive amounts of low-speed torque coupled with scintillating acceleration in the upper reaches of its engine speed range. To achieve this state took more than a year of intense development by Power Engineering, whilst a new crankshaft and piston assembly was designed and made. Machined from a single billet of high-grade steel, the crankshaft differed greatly from that of the standard engine in that it featured small, balanced pairs of counter-weights to each conrod, rather than one big one, thus enabling the engine to spin more freely with a minimum of inertia to overcome. Forged pistons and race-grade conrods completed the bottom end of the engine.

The cylinder head, camshaft profiles and turbocharger installation were also uprated and the management system remapped to allow maximum benefit to be derived from the complete package. The 2.4 engine has the driveability of a small V8 and the top-end surge of a Phantom jet on full afterburn.

To cope with these changes, the chassis needed uprating, which was where P&P Performance became more involved with the project. Mark Lovell's expertise as a rally driver is equalled by his ability to set cars up for events, and both attributes were utilised to result in a clean-handling, precise and controllable machine. The first car was a 4x4 version, but subsequently a two-wheel-drive Sapphire was also modified, requiring quite different suspension settings to cope with the different behaviour patterns of the rear-drive car. Both packages have since become available for drivers seeking to raise the performance characteristics of their cars and at the same time to improve driveability.

Above: The green cam cover on the RS2400 is an allusion to its ability to run happily on lead-free fuel. The grille badge has been picked out in the same colour.

Previous page: The Power Engineering/P & P Performance RS2400 gets its name from the 2.4-litre engine conversion which is the heart of the package. This gives much-needed low-speed torque with further gains in on-boost response.

In terms of outright acceleration, the car can run in four-wheel-drive form from standstill to 60mph in 4.65 seconds, clipping a solid two seconds from the standard car's time. More significant, though, is its ability to halve the time needed to overtake: the machine takes a mere 4.5 seconds to accelerate from 60 to 90mph in fourth gear. Outright speed is also increased, the car proving itself on a long run in France, with the author at the wheel, of hitting 165mph.

In the author's exclusive first road test of the car, the 4x4 was described as an automotive cross between Arnold Schwarzenegger and Fred Astaire — the combination of lightness of balance and sheer muscle-power was a heady and addictive one. Subsequent encounters have reinforced that view.

In two-wheel-drive form, the RS2400 is more of a handful — there is the power instantly available to punt the back of the car out of line on bends — but it, too, is a stimulating and rewarding experience.

The Wide And Wild One

Frank Cirino's car started life as one of the original 5,000 RS Sierra Cosworths and was two years old when he bought it. Like many owners, he ran it as standard for a while and was generally happy with it, but then became increasingly desirous of a higher level of performance.

The engine in the car is now a 'Desert Storm' powerplant, this being the brand name Turbo Systems of Staffordshire use for their full-house engine conversion. A nett power output of 500bhp was achieved by altering almost all of the induction system, making use of a special hybrid turbocharger designed and built by Turbo Technics. To handle these changes, a steel billet-machined crankshaft, Cosworth 7.2:1 pistons and a host of other block changes were made to what began as a YBD engine from an RS500.

Those wide arches — which are an adaptation of a Thundersaloon's styling package — are there to cover a monstrous set of wheels, the fronts measuring 16" x 10" with 255/40ZR tyres whilst the rears are 17" x 12" with 335/30ZRs. These brought with them the need

Below: *Frank Cirino's RS Cosworth is the result of a massive investment programme — £60,000 has been mentioned — and features a full-specification 'Desert Storm' engine by Turbo Systems to move those massive-section rear tyres.*

Left: Body styling is based on the package developed for Thunder-saloon racing and was a requirement of having such monstrously wide tyres on each corner of the car.
Below: Ferrari-esque rear haunches were a major headache to design and build — aligning the 'blades' of the side strakes proved extremely difficult. But the result was worthwhile, making the car the widest Cossie on the roads of Britain.

to totally revise the suspension system, which now has specially-valved Koni coil-over-damper units, high-poundage springs and special lower links and bushings. To handle the traction lock-up, a Quaife 4.43:1 epicyclic differential assembly features. The lowered final drive ratio is necessary to compensate for the much greater rolling radius of the fat rear tyres.

Inside those huge wheels are race-grade brakes, essential to handle the power and traction available from the much-altered car. To give an idea of just what the term 'race-grade' means, the rear discs are bigger than the front discs on a standard RS Cosworth, whilst the fronts are a massive 330mm diameter and 30mm thick. These items, manufactured by the Italian company Tar.Ox, are squeezed by AP Racing callipers.

To achieve the car you see here, Frank Cirino spent a total in excess of £60,000, making it possibly the most expensive Cosworth in private hands.

The Land Speed Record Cosworth

Above: Mike Lansdell (in car) and Charles Eveson carry out a familiarity check prior to making the first of two timed runs in opposite directions.

Land speed record attempts are rather rare these days, especially in the wheel-driven category. The lure of jet- and rocket-engined machines, with their 600+mph capabilities, is what engages the attentions of latterday Campbells, Cobbs and Segraves.

However, there is one current land speed record which is held by a Sierra Sapphire Cosworth 4x4 — that established at Pendine Sands, South Wales, in September 1991. With an average speed over two passes of 130.75mph, the record might not sound like much to write home about; the Cosworth is, after all, eminently capable of sustaining such a speed all afternoon on the German *autobahnen*. But add to the equation the fact that the driver is blind, and the achievement is lifted into an altogether different perspective.

Michael Lansdell was the driver responsible for setting the record, which involved a flying quarter-mile pass in two opposite directions along the Pendine beach. (A 45mph headwind on the return run cut the speed from a first run time of 147mph.) Lansdell is officially classified as Grey Blind, which means that he can detect silhouettes but nothing more.

To be sure of aiming the car directly, he needed a co-driver, of course. Charles Eveson, principal of Hartford Motors, a Ford dealership in Oxford, took on this role. Acting as Mike's eyes — but definitely hands off the controls — Charles was able to ensure that the car was perfect-aimed into and then along the RACMSA-laid course.

The Sapphire 4x4 used for the exercise, also supplied by Hartford Motors, was surprisingly close to standard. A BBR Racing management system and actuator uprate increasing the power output to about 285bhp and a special set of high-grip Goodyear 205/50 tyres were the only changes made to the car. As a testament to its versatility, Charles Eveson had driven the car to Pendine prior to the event and then drove it home again afterwards.

BEN, the charity for which Mike Lansdell has worked ever since losing his sight, has erected a commemorative display board at Beaulieu's National Motor Museum, as a permanent tribute to his record. And his photograph hangs alongside those of the other famous record holders, adorning the walls of the Pendine Hotel.

Left: The bleak expanse of Pendine Sands, South Wales, was the venue for 1991's Blind Driving Land Speed Record. Using a Sapphire Cosworth 4x4 supplied by Hartford Motors and tuned by Brodie Brittain Racing, Mike Lansdell set a new record of 130.75mph and a permanent place in the Pendine 'hall of fame'.

Below: The off. Special tyres were provided for the attempt by Goodyear, which would improve traction on the sandy surface. Brown Brothers Dana, for whom Mike worked before his BEN appointment, provided logistical support for the event.

The Sierra Cosworth Engine

The engine which powers the RS Sierra family is based on the cylinder block of the rest of the old Sierra range, known as the Pinto block after the American sub-compact car in which it first appeared.

Under the generic heading of YB, the engine is of over-square dimensions, the bore of 90.8mm being appreciably greater than its 77mm stroke — such a dimension is accepted as the ideal way of getting what is essentially a racing engine to rev freely to a very high speed. That the block dimensions were carried straight over from the donor engine was a bonus for Cosworth, in that they were able to take the complete block and crankshaft assembly and build on it.

New pistons which lowered the compression ratio to a manageable 8:1 were created and the lubrication system was upgraded to ensure adequate lubrication (later for the RS500 and the 4x4 the oil system was further improved), with under-piston sprays of oil acting to cool these components. The crankshaft and conrods were heat-treated to ensure longevity, but in all other aspects the T88 engine block was just like that of a 2.0i Ghia.

Not so the head. This was a wonderful feat of pressure-cast engineering achieved in-house by Cosworth, with their familiar pent-roof chamber design featuring a pair of 35mm inlet valves on one side and a pair of sodium-filled 31mm exhaust valves on the other. The single spark plug was situated slap-bang in the middle of the quartet

Below: The twin camshaft engine with Garrett turbocharger fits neatly into the Sierra engine bay. The cover plate for the turbine unit is essential to guard against the bonnet paintwork scorching.

of valves, in classic Cosworth style. A T.03 Garrett AiResearch turbine unit sitting on a specially-developed intake manifold fed the chambers, which gained their fuel from injectors mounted as close as possible to the intake throat. An electronically- controlled wastegate dumped any turbine pressure beyond the predetermined 8lbs.

A management system by Weber Marelli, the electronic fuel injection specialists, controlled mapping of both fuel and ignition, the latter being timed by a crankshaft sensor. Much work went into the development of this system and into its 'soft cut' engine speed-limiting feature, which again was intended to ensure a long and happy engine life.

Interestingly, whilst the engine was first developed as a private venture by a team at Cosworth and later picked up on by Ford management in need of a suitable powerplant for their road-racer, it was never intended by Cosworth to be a turbocharged unit: they were looking for a consistent 200bhp in naturally-aspirated form when Ford interrupted and changed the game plan. In 1991, when turbochargers were outlawed from Touring Car races, a number of racers were forced to look seriously at the engine in its original form — and results bear out Ford's feeling that the turbine unit was essential to extract sufficient power from the engine.

That first non-turbo engine was known within Cosworth as the YBA, and its forced-induction successor as the TBB. Further variations on the theme were to appear, although not all went into production cars: the YBC was a Mountune-prepared rally engine, the YBD was the evolution engine for the RS500, the YBE was a special variant for non-Ford applications such as the aborted Panther Solo project, the YBF was a racing version of the RS500 unit, the YBG was the lead-free version of the YBB engine which met 83US emission standards, the YBJ was the variant used in the 4x4 Cosworth, and the YBP was the version able to meet all confirmed 1993 European emission regulations. Finally there is the YBT powerplant, which has the new hybrid T.03B/T.04 turbocharger and a Ford blue camshaft cover. It is this engine which will be used in the RS Escort Cosworth and in the Ghia Focus sportscar likely to appear in 1994.

In each case the engine number is preceded by the appropriate engine type designation, making identification a simple task.

Above: The RS500 unit boasted an additional four injectors — not connected on road-going cars — which were on a separate fuel rail mounted above the original intake manifold. Bigger air trunking led to a suitably bigger T.04 turbocharger.

Below: The complicated manifolding used on the Cosworth powerplant is designed to maximise response time for the turbocharger. The iron castings are bulky, but offer good gas-flow characteristics aligned to support for the turbo unit.

The Sierra RS Cosworth In Motorsport

The entire purpose of the original Sierra RS Cosworth was to be a racing car: road legal versions were produced only to satisfy the requirements of motorsport's governing body. The necessary approval is known as homologation, and the rule at the time of the car's creation was that there must be at least 5,000 examples built — in other words, the model must have a substantial production run, rather than just being a 'few-off' special.

In both racing and rallying at international level there is a considerable degree of overlap between the cars, not least that both branches of motorsport have 'showroom' classes (Group N) and 'evolution' groups (Group A). For Group A, the classification is for 10 percent of the initial run.

By getting on to both lists, the Sierra Cosworth was immediately

Above: Group A Touring Car Racing is a big-money business, as this shot of the Graham Goode equipe shows. As well as two race-cars, the team also uses a 40-tonne articulated support vehicles (with side canopy) and a couple of pickup trucks. Budgets can run to half a million pounds per season.

eligible for both racing and rallying.

Early on in the development programme which led to the Sierra Cosworth, Coventry-based saloon car racer Andy Rouse had come into the picture, and throughout 1985 and 1986 his Merkur XR4Ti (a 2.3-litre variant of the XR4i built for the American market) dominated the top end of British saloon car racing. The lessons learned by Rouse and his experienced team found their way back to Ford and were integrated into the Sierra RS Cosworth programme. When the Cosworth debuted in the British Touring Car Championships in 1987, there was Rouse, in a Ford Motorcraft-sponsored example of the whale-tailed beast.

A year later, Rouse moved up to the RS500. (The car had been homologated for August 1 1987 and so did not make a serious impact in British racing until the following season.) The overall 1988 series crown went to Frank Sytner, whose BMW had scored more consistently in the class win points, but Rouse and his team were Class A winners. As well as the Rouse Motorsport car, there were also RS500s being piloted by a number of other teams, with such rising stars as Tim Harvey and Gerry Mahoney notching up successes.

Meanwhile, the 'ordinary' Sierra Cosworth was dominating the production car scene. The Castrol-sponsored entry of Sean Brown dominated the Monroe Saloon Car Championships, fending off threats from a dozen other Group N Cosworths as well as the rival marques lined up on the grid.

Success was also achieved in mainland Europe. The Texaco-liveried RS500s of the Eggenburger team stole all honours in the World Touring Car Championship in 1987 — the only year of the series — and went on to countless successes

Above: One of Britain's most respected constructors is Andy Rouse. The ultra-tidy under-bonnet of his highly successful RS Cosworth gives an idea of how much care and attention goes into a typical Rouse product.

Below: The adoption of brake ducting pipework for the RS Cosworth was essential to provide adequate cooling air for what would in a race become red-hot pieces of metal. On some circuits the cars might be expected to slow from 170mph to 60mph in only 200 metres.

Left: *Rather than side glass, many drivers have opted to use netting first intended for use in NASCAR racing in America. This is highly effective at containing the driver in the event of a collision.*

Below: *Racing in the streets. The centre of Birmingham was turned into a race-track over August Bank Holiday weekend for several years in the late eighties — and the Touring Cars were a highspot of the racing card. Here Andy Rouse lies third, behind arch-rival Robb Gravett and leader Tim Harvey.*

in the German Touring Car Championships. In Australia it was Dick Johnson who took the RS500 to a succession of wins in the country's equivalent series. His winning car from the 1988 season came back to Britain in the winter of that year, having been purchased by Rob Gravett for the fledgling Trakstar operation. At the end of 1989, Gravett, in his first year of Group A racing, was on top of the championship, having beaten off sustained efforts from both the Rouse-Kaliber team and the Bristow-Labatts entries to take the ultimate honours.

As Stuart Turner of Ford Motorsport had already observed, the RS Cosworth package was, if anything, too successful By beating everything in sight, the cars effectively knocked all the spirit out of the other big players in the touring car race scene, with the result that FISA was left in a position to do nothing but cancel the Cosworth's eligibility for Group A racing. In 1990, the RS500 was still able to take part in Esso BTCC rounds, but only in an emasculated form: the restrictions imposed on the induction system were sufficient to neuter the car and it was never to win another Group A race.

Ironically, whilst the RS500 saw off all opposition on the racetracks, it was never anything like as successful in rallying events: the early cars suffered from a lack of traction and the later ones from

Above: *Almost a legend, Dave Brodie of BBR has never been one for quiet colour schemes: this is his chrome-sided 1989 car. The RS500 made only a handful of appearances during the season, but Brodie's driving style thrilled all who watched him.*

Above: The awesome budgets required by Touring Car race teams were becoming increasingly difficult to find — until brewers Guinness decided that the formula was an ideal platform for their alcohol-free lager, Kaliber. Andy Rouse was the recipient and did them proud with a string of wins.

unreliable engines. In the International series, where Ford had shown the way in the 1970s with a succession of Escorts, the Group A RS500 was able to produce only one win, that being on the Tour De Corse in 1988 with Didier Auriol at the wheel. That solitary success apart, the best Ford could manage during the car's rallying career was a string of lesser placings.

Even after the car gained the benefit of four-wheel drive halfway through the 1990 season, matters failed to improve for Ford's Group A rally challenge. Despite some formidable driving talent — Pentti Airikkala, Alex Fiorio and Malcolm Wilson, as well as the great French hope, Francois Delecour — the works cars failed to win a single major event. Up to the time of writing, 1992's second place on the Portuguese rally was their best result.

Fortunately, Ford's corporate face has been saved throughout the period by the less glamorous 'showroom class' cars which run in Group N. These have consistently won their class, notching up important points for Ford in the league table as they did so. As well as a good number of factory-supported teams, the Cosworth in Group N form has been a strong favourite with privateers, who have capitalised on its blend of high power, great traction (in 4x4 form) and inexpensive performance equipment.

At the point of completing this book, the next chapter in the Ford competition story is about to be written. The replacement for the Sierra Cosworth, a Cosworth-engined four-wheel-drive variant of the new Escort, is waiting in the wings. It is due to make its international rallying debut as a works car in 1993, though there is a possibility that one or two will appear in the 1992 RAC Rally. With the Escort, Ford intend to restore themselves to the top of the leader-board in rallying — a place which they almost justifiably feel to be theirs by right — and to make the Escort name once more synonymous with success.

The new car draws heavily on the technology of the Sierras which are the subject of this book, and it is fair to say that without the Sierra programme, the new Escort programme would not exist. And therein lies an irony, since without the unsuitability of the 1980-series Escorts, the competition Sierra would not have existed either…

Below: Former helicopter pilot Lawrence Bristow became a team owner in 1987, with more than a little help from Labatt, the Canadian brewery. Rumours that he would have to equip the team with Mountie hats in order to tie in with current Labatt advertising were unfounded…

Left: *Team-mate to Bristow was the exceptionally talented Tim Harvey, who despite suffering from tyre compound and durability problems throughout the 1989 season, made his mark with a series of stunningly effective drives. The affable Harvey has moved over to BMW since then, but continues a Cosworth connection in Group C endurance racing.*

Below and opposite page, below: *Despite the combination of ability on both his and TV star/team-mate Mike Smith's part, and a superb level of presentation, the Trakstar team of Robb Gravett failed to attract sponsorship until mid-season and thus both cars ran for much of '89 in plain white.*

Right: Throughout the golden age of the Sierra Cosworth in Group A racing, the model achieved total domination, providing a level of action and excitement which is unlikely ever to be seen again. Once again, this is Rouse (back in ICS colours), Harvey and Gravett heading the pack at Silverstone in 1990.

Below: As the rules changed, Rouse Racing began developing an atmospheric contender, using a Sapphire bodyshell with non-turbo'd Cosworth engine. In the hands of Ray Bellm, the WSC driver, the car was moderately successful but ultimately unable to beat the BMW M3 entries in Group A Class Two.

Above and below: Brodie Brittain Racing's answer to the atmo-two-litre rule was to go to four-wheel drive — this would offer increased traction and reduced tyre wear. Unfortunately, the formula was not given sufficient time to come good. Subsequently sold, converted to rear-wheel drive and equipped with a Group A turbo RS500 engine, the car now races in Thundersaloons.

Above: *For 1991, Gravett moved to an atmospheric Sapphire Cosworth, in Shell livery. Here he attempts — but fails — to head the unassuming Andy Middlehurst, who had moved up from saloon car racing to drive the sole Graham Goode Racing atmospheric Cosworth hatchback.*

Left: *Not quite a wedding car… but almost. This was Graham Goode Racing's acknowledgment that Andy Middlehurst had interrupted his honeymoon to take part in a round of the Esso Touring Car Championship in 1991!*

Below: *Those who served. Some of the top guns in Touring Cars who drove RS500s during the golden years. From left: Andy Rouse, Lawrence Bristow, Robb Gravett, Tim Harvey, Mike Newman, Graham Goode, Sean Walker.*

Right: The RS500's days in motorsport were continued into 1991, with the Collins Engineering machine in the hands of journalist, author and racing driver Jeremy Walton. With 540bhp and street tyres, the car was a highlight of the RS Owners' Club-inspired Vecta Challenge… When this author tried the car at Oulton Park, the only printable adjective was 'awesome'.

Below: In Saloon Car Racing, which is run effectively to production car levels of performance, the Cosworth continued to dominate. This group is seen in Britain's only 24-hour race, the Willhire, at Snetterton.

Right: The Willhire 24 Hours gives the car a severe test — and provides plenty of exercise for the pit crew who have to refuel, re-tyre and re-brake the cars every ninety minutes during the event.

Above: The only car which could cut it with the Cosworths during the Esso Saloon Car Series was the Saab Turbo of Ed and Lionel Abbott. They had moved over to the marque after a great deal of success with the Cosworth in the previous season. Here Steve Monk heads off their challenge at Silverstone.

Right: Two generations together: the Sapphire of Steve Monk is headed by the vivid purple hatchback of Graham Davis. The Esso Saloon Car Series was the only production car series where such a battle would be seen by 1991.

Right: Race-car constructor Kevin Maxted moved neatly into the 'rent a racer' market with the Cosworth. Here one of his fleet of powder-blue machines is being used by former rally driver (now Ford main dealer) Peter Clarke, from Skipton.

Above and opposite page: *When the RS Cosworth was introduced, Ford had their studio produce a group of illustrations of the proposed liveries. These were Ford's own Motorsport colour scheme, the Texaco livery, British Telecom Radiopaging, and the French '33' Export. All four were used during the car's first year in rallying.*

Above: *Anatomy of a 'nearly' car… Despite substantial resources and hard cash being pumped into the works rally team, the Sierra Cosworth never achieved the success which was hoped for.*

Opposite page: *For 1986 Ford took seven of the cars from the initial Sierra Cosworth press launch and converted them for rallying. A unique championship, sponsored by Securicor Communications, was run for 'The Magnificent Seven', using five of the National Rally Championship rounds. Loaning the cars to seven of its dealers, Ford was thus able to test the Group A potential of its new machine, while running the Group B RS200 at works level.*

Phil Collins, the favourite, eventually took the Securicor Sierra Challenge in the Brooklyn Garages car, but only just. He came to the final round, the Lindisfarne Rally, tied on points with Chris Mellors. Mellors led first, then Collins; Mellors attacked back, but Collins was equal to the challenge. At the end he was 10 seconds — and one point — ahead of Mellors. Securicor went on to enter the top three in the Challenge, Collins, Mellors and (seen here in the Kielder Forest during that exciting last round) Rob Stoneman, in that year's RAC Rally. All three acquitted themselves well, but retired at roughly the same time. (Ian Wagstaff)

Above: *Amiable top-seeded rally ace Pentti Airikkala won the 1989 RAC Rally in a Mitsubishi and thus qualified for the coveted Number One door plate for his 1990 entry in the same event. A works entry with finance by Autoglass almost came good, until a suspension break on the third day of the gruelling event brought his rally to a premature end.*

Left: *The braking system of a Group A rally car is every bit as important as the engine. This rear disc assembly shows just what stopping power is available to the driver — ventilated discs with cooling grooves and four-pot AP calipers can be controlled by a bias valve on the fascia.*

Right: The office of a rally car in Group A trim is a complicated affair. Ultra-accurate time and distance computing equipment for the navigator occupies the passenger side, whilst purely functional instrumentation surrounds the driver.

Below: Every car has its own support crew, in a fully-equipped service barge which carries a full complement of spares. Backing this up will be a flotilla of other vehicles, and usually a tracker helicopter as well. An event such as the RAC Rally will cost £5 million or more per works team.

Right: At the opposite end of the scale from the major players are the privateers, who will do everything in their power to win — often relying heavily on accommodating family and friends to augment the budget.

Below: The Cosworth was responsible for the return — just for the one event — of one of Britain's top rallying stars, Roger Clark. The event was the 1989 Autoglass Tour of Britain and Roger was drafted in at the last minute. A new firesuit was rapidly obtained when he found that he could no longer fit in the one he had hung up earlier in the decade!

Left: *Francois Delecour was recruited to the Ford of Britain works team in 1990 and immediately attacked the challenge of making the Cosworth a workable proposition. Outright success eluded him, but his driving skills made the most of what one of his team-mates later described in an aside as a 'third world car'.*

Left: Ford's fortunes in Group A have been compensated for by some fine results in the showroom Group N class. Louise Aitken-Walker, one of Scotland's finest drivers ever, joined the team in 1991 and immediately scored a class win on the RAC Rally.

Above: One of the most successful of privateers, Robbie Head, is seen here attacking the pack in his 1991 Shell Oils-sponsored Sapphire 4x4.

The Cosworth in Miniature

Cosworth owners are of course a discerning group of people, and accordingly there is a vast range of after-market accessories, from a variety of sources, aimed fairly and squarely at them.

As well as an abundance of leisurewear, encouraging owners to proclaim their allegiance when away from their cars, there are also a number of scale-model Cosworths available. Most of them are high-quality diecasts from British or German manufacturers such as Schabak or Motor-Pro, and often represent the cars in fine detail. The Motor-Pro range majors on liveried replicas of sporting variants, whilst the Schabak choice is between the road-going Sapphire 4x4 and the same car in rallying livery.

The model which has most fired the imagination of Cosworth owners, though, is the radio-control 1/10th scale RS500 released in 1991 by the Kyosho Corporation. This is a superbly accurate model of the German Touring Car version, raced with stunning success by Klaus Ludwig and his colleagues for Rudi Eggenburger.

As an alternative to the Texaco livery, enthusiasts can also buy a decal set for the ICS Andy Rouse team car or the Listerine/Graham Goode Racing car, both of which ran in the British Touring Car Championships. The author is himself heavily into scale-model race-cars; he owns the Kyosho RS500 shown here.

Below and opposite page, below left: The most sought-after replica of the Cosworth is the Kyosho RS500. The author has two examples of these one-tenth scale radio-controlled cars (imported into the UK by Ripmax plc), both featuring custom paint schemes.

Top and above: *These two rally-specification Cosworths — the Sapphire of Russell Brookes from 1991 and the 1988 hatch of Carlos Sainz — are by Motor-Pro, whilst the road-going 2WD Sapphire, in the bigger 1/24 scale, is by Schabak*

Inheritors: Beyond The Sierra

Although the Sierra Cosworth story ends in April 1992 with the arrival of the new CDW-27 Sierra replacement, the name will not die off. There will be a turbocharged, 200bhp, Zeta-engined car with front-wheel drive — although a 4x4 chassis is also a distinct possibility.

As for the tremendous amount of research and development which went into the Sapphire 4x4 project, this will not be wasted because there is one definite car, and another nearly definite, which will make full use of all that technology. The definite is the Escort RS Cosworth, utilising a modified Sapphire 4x4 floorpan in conjunction with the familiar MT-75 gearbox and four-wheel-drive system, wrapped up in a suitably wide-arched Escort CE-14 bodyshell.

The 'nearly' car is the new Ghia Focus, an aggressively-styled sportscar which debuted at the Turin Auto Show in April 1992. This car again makes use of the same floorpan (although the wheelbase is reduced by 100mm) and 227bhp Cosworth YBT engine with four-wheel-drive system. Introduced as a concept vehicle, the Focus is a distinctly feasible limited-production car — indeed, that Ford have said as much in their launch literature leads one to believe that a production

Below: The new RS Cosworth Escort is set to take over from the Sierra RS as the prime works-entered competition machine. A further refinement of the engine and drivetrain assembly seen in the Sapphire RS 4x4 underpins a radically-altered CE-14 Escort bodyshell.

Opposite page, above: A distinct prospect on the horizon is the Focus, a Ghia styling exercise which was shown as a fully-working prototype early in 1992. The car may become a production reality by late 1993. Once again the Sapphire RS Cosworth 4x4 running gear (albeit with the latest YDT catalysed engine) is the base for the dramatically styled machine.

Below: Totally different to anything you might have seen inside a Cosworth, the Focus interior features saddle leather seating, 'floating' instrument cluster, and a wood veneered floor with appropriately-placed grip pads. Dashtop is brushed stainless steel as is capping of central tunnel and door armrest.

version is not that far away.

Whatever the future, one thing is certain: there will never be another Ford road car which will be quite as successful on the race-tracks of the world as the Sierra Cosworth has been.

For those with a hankering for a 'real' Cosworth, BBR, the road and race-car equipe headed by racing driver David Brodie, are now re-manufacturing the RS500, taking new Motorsport bodyshells and building them up into new cars by hand. At the time of writing, the retail price is a cool £59,995…

Ian Wagstaff

THE AUTHOR:

A journalist who also takes photographs, **Dennis Foy** *(left) is founding editor of* Performance Ford *magazine. He has possibly driven more Sierra Cosworths than any other motoring journalist, having road-tested not only all the standard models but many uprated versions. His record so far stands at 590bhp, but he feels there is potential as yet untapped. When not working, he lives in Cheshire with his wife Pat and son Ben, along with their four cats. His previous books include* Escort Performance, Ford Escort *and* Automotive Glassfibre.

THE PHOTOGRAPHER:

Terry West *is the motorsport photographer in the* Performance Ford *team and is normally to be found lying prone alongside one of Europe's race-tracks, aiming a substantial Canon camera at fast-approaching racing saloons. Other publications in which his work has appeared include* Automobile Quarterly (USA), Ferrari World *and the prestigious motor racing annual,* Autocourse. *He lives in Surrey with his wife Jackie and two children.*